The Other Plants

Small Lessons about Some Non-Flowering Plants

A Coloring Book

By Michael Reed

The Other Plants

Small Lessons about Some Non-Flowering Plants

A Coloring Book

By Michael Reed

Copyright©2018 by MR

All Rights Reserved

Acknowledgment

I thank God for giving me the knowledge and interest for this book.

Non-flowering plants are an important part of our world because they helped to contribute to our lives and the environment.

A tulip is a flowering plant because its seeds are enclosed in an ovary.

Like the tulip, the sunflower is a flowering plant.

Non-flowering plants are diverse in many ways.

Here are a few examples of some of these plants on Earth.

Liverwort

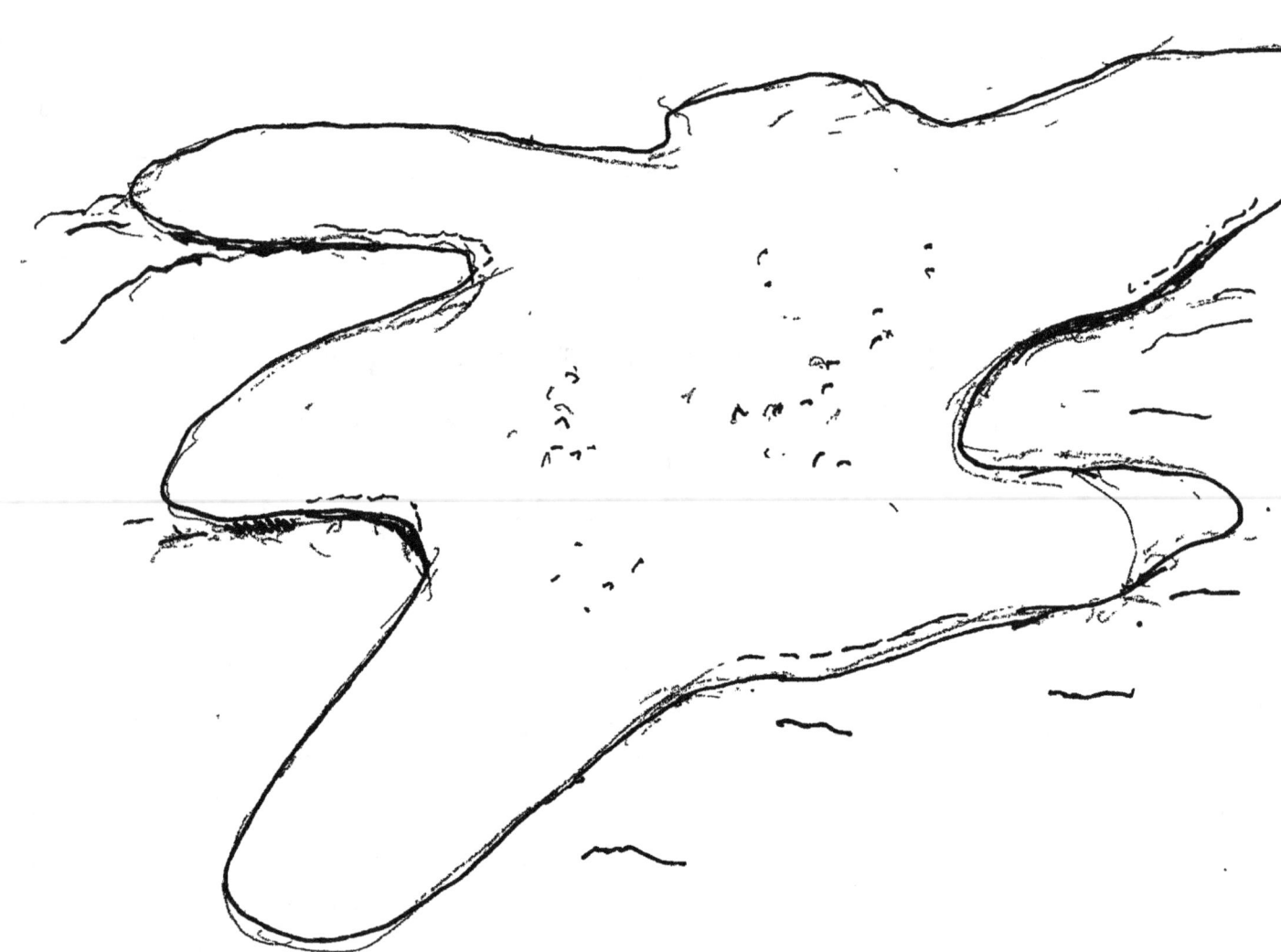

Tree fern

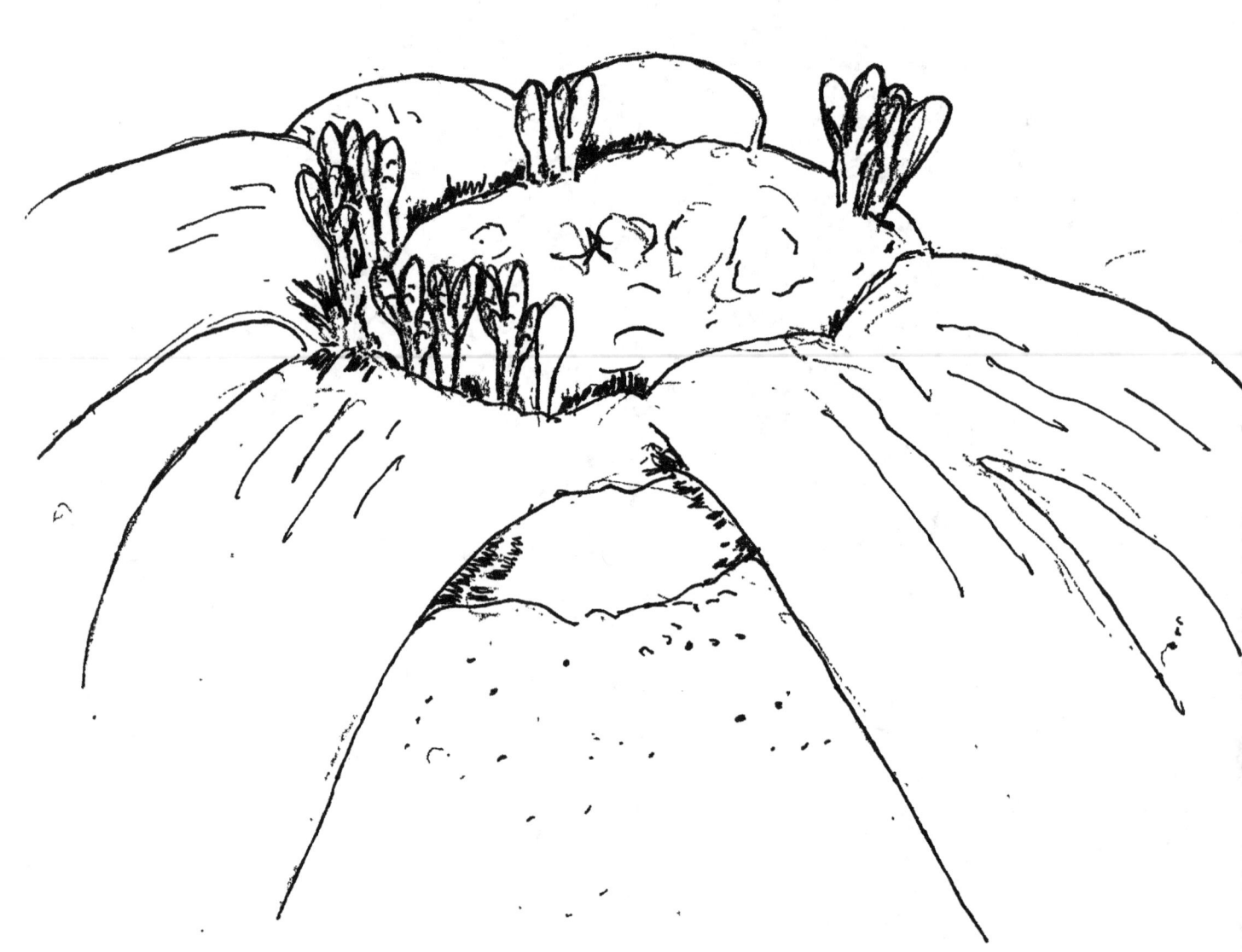

Welwitschia mirabilis, a plant that is native to the deserts of Southwest Africa.

Horsetails

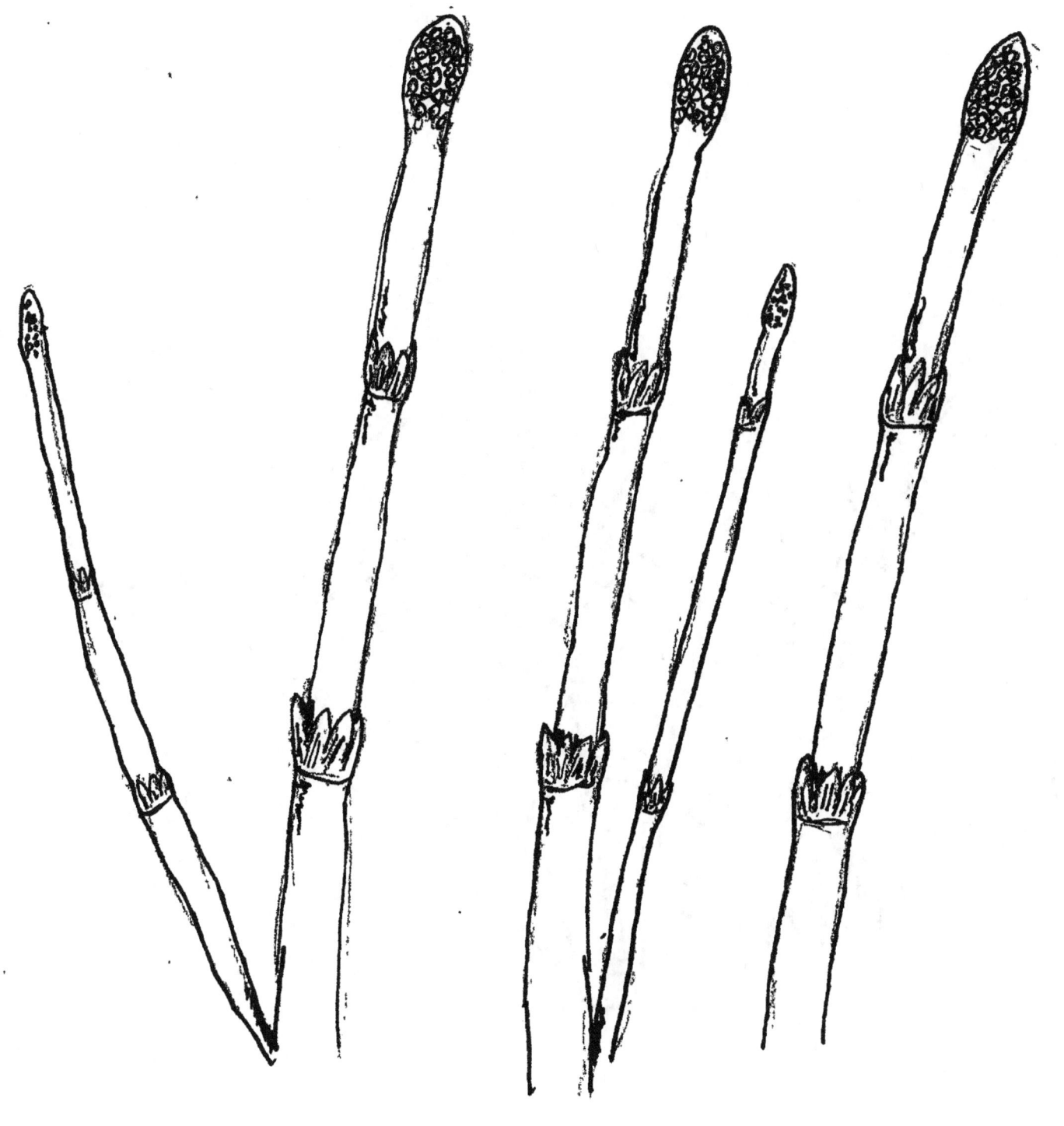

Fern

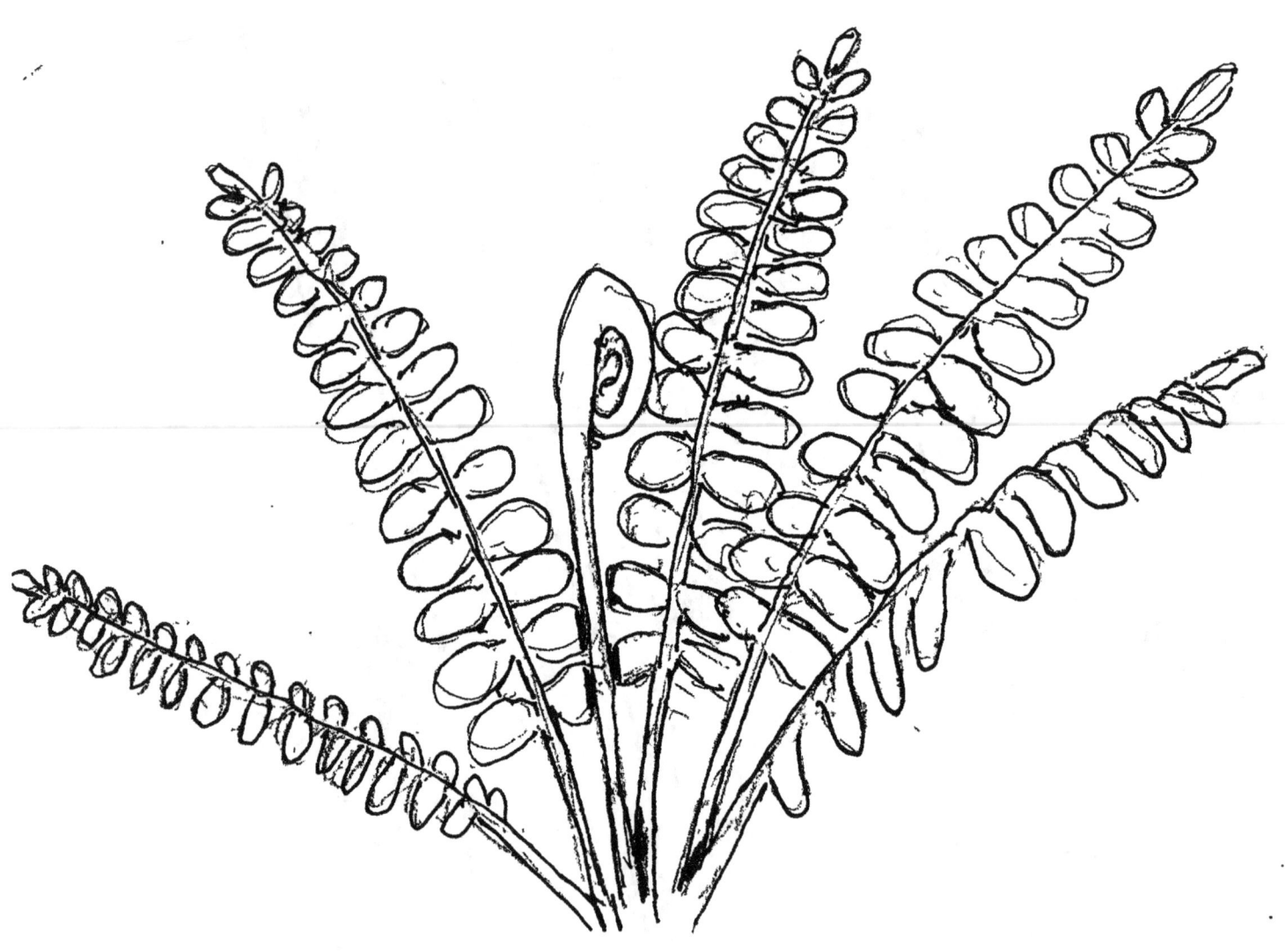

Each plant has various ways of reproduction.

A yew tree has its seeds enclosed in a fleshy covering called an aril.

A pine has exposed seeds in its cone.

A fern has its spores in capsules.

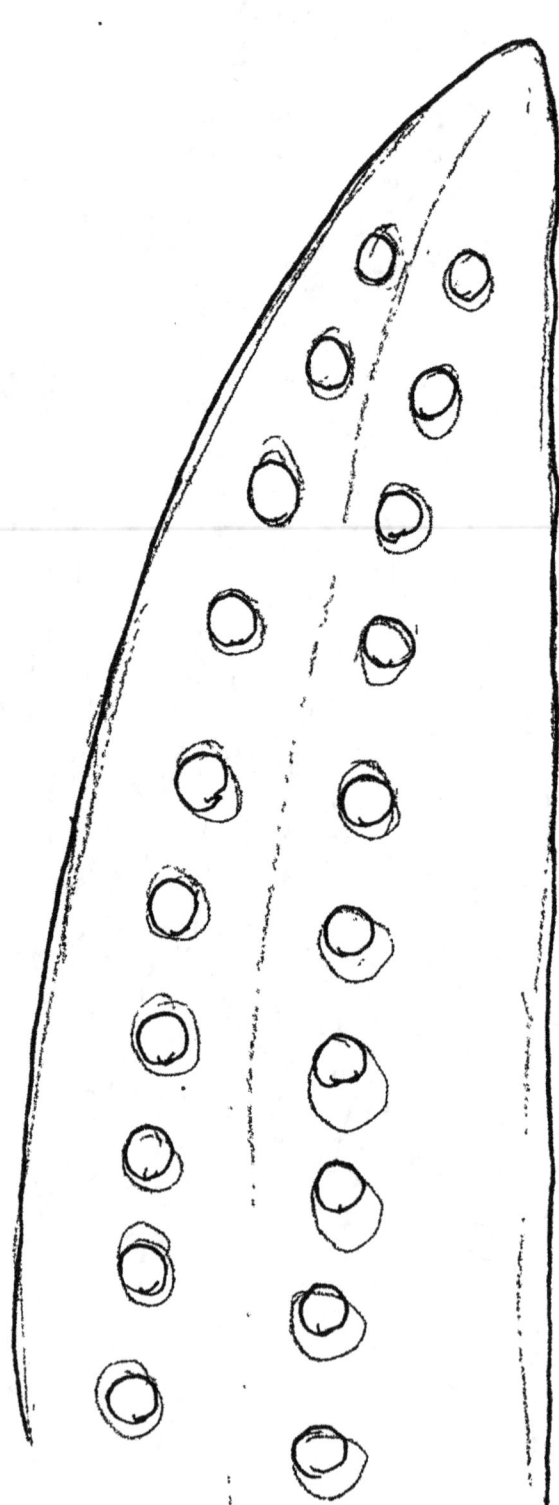

A Life Cycle of a Fern

A prothallus is grown from a spore.

A fern releases its spores from their capsules.

Once the male and female cells combined, a fern grows in its right conditions.

Mosses release spores from their capsules.

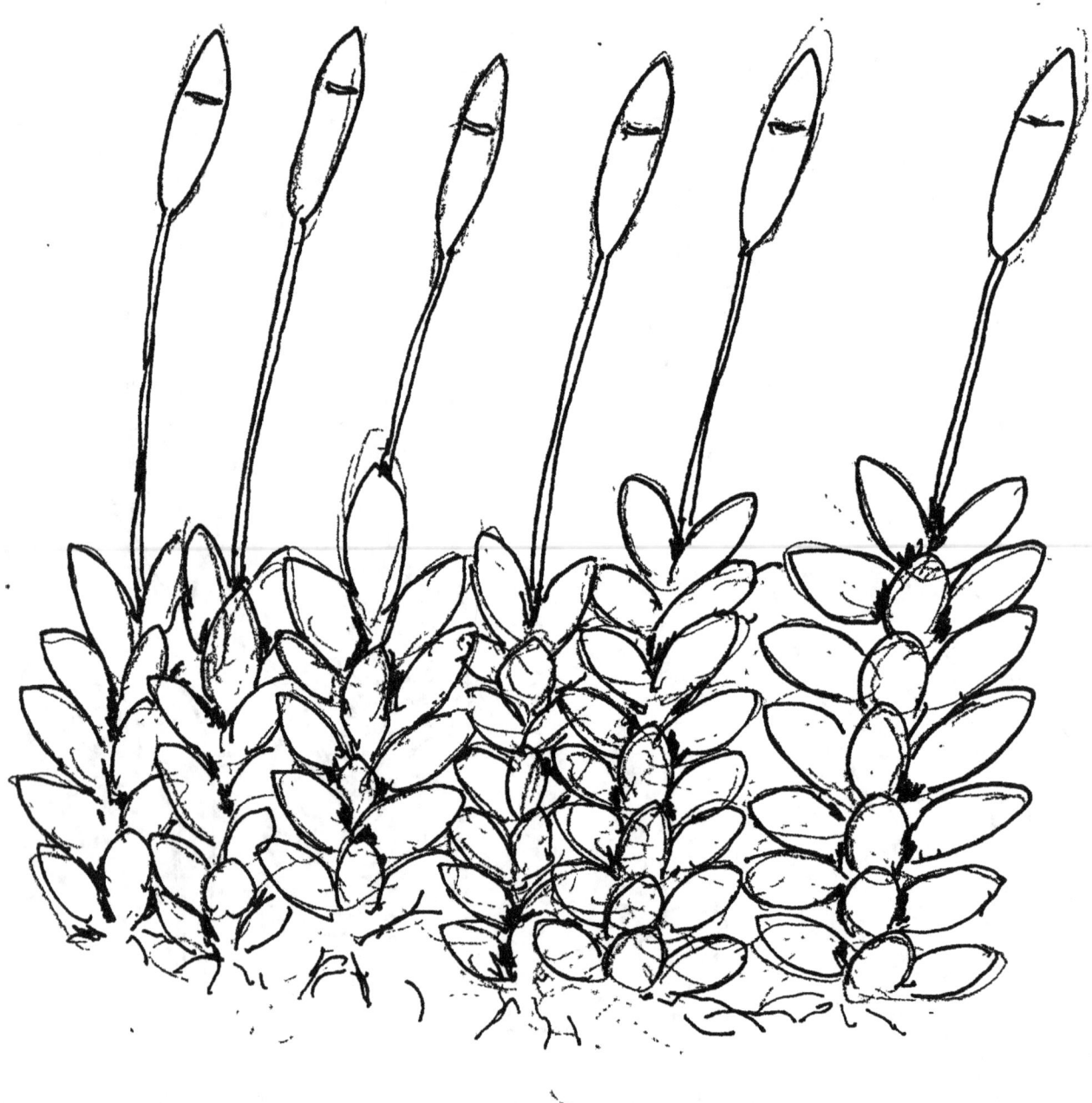

We use many non-flowering plants as...

... food

The seeds of some pine species are used as food.

The fiddleheads, uncurled leaves, of some fern species are used as food. However, caution must be taken with several species.

...medicine

Ginkgo biloba is used as medicine.

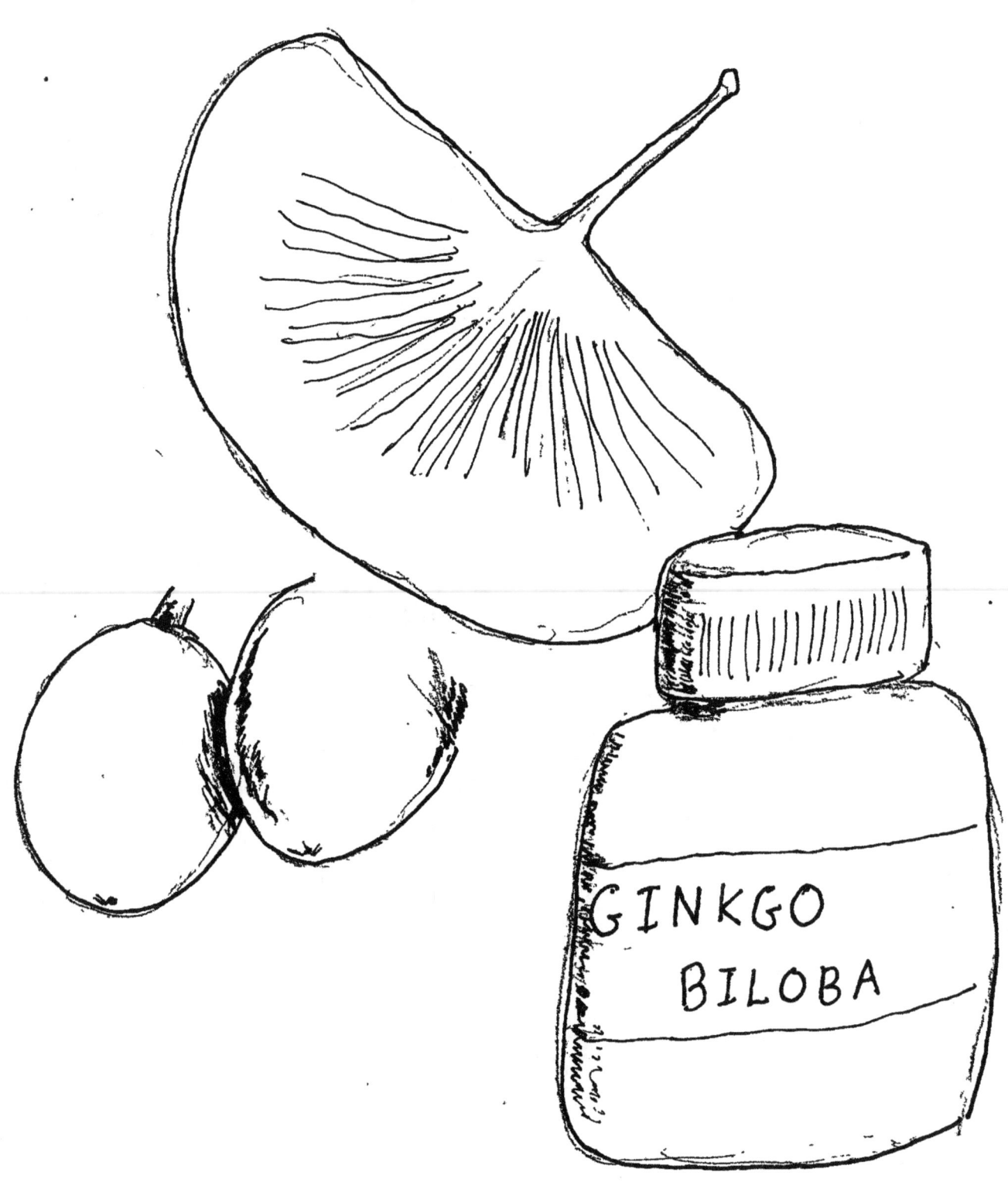

... and decoration.

Several conifer species are used as timber in making of houses.

Many kinds of non-flowering plants are an important contribution to various ecosystems on Earth by helping to provide nutrients and cover for other plants and habitats for animals.

A Redwood (Sequoia sempervirens) Forest in the Pacific Northwest region of North America.

Tree ferns in a tropical cloud forest.

A Bald Cypress (Taxodium distichum) tree in a swamp of the Southeastern USA .

We can learn more about non-flowering plants through some museums, books, teachers, and the computer. However, we can turn to God to get our understanding about them.

References for Use

Cullina, William. **Native Ferns, Moss & Grasses**. Houghton Mifflin Company. Boston. 2008.

Field Museum of Natural History. Chicago, IL

Mauseth, James D. Botany: **An Introduction to Plant Biology**. 6th Ed. Jones & Bartlett Learning; Burlington. 2017.

Parker, Steve. **Kingdom Classification: Ferns, Mosses, & Other Spore-prodcuing Plants.** Compass Point Books; Minneapolis. 2010.

www.ingramcontent.com/pod-product-compliance
Lightning Source LLC
Chambersburg PA
CBHW062236220526
45471CB00009B/3508